Savannahs are lands covered by tall grasses, scattered trees, and thorny bushes. They are also known as tropical grasslands. You can find large stretches of savannah in parts of Asia, Australia, and South America. But the largest savannahs are in Africa. They cover nearly half the continent.

It is always either warm or hot on the savannah. The seasons are marked by wet weather and dry weather. During the summer months there is lots of heavy rainfall. Green grass sprouts and the scattered trees grow leaves during this wet season.

Little or no rain falls during the rest of the year. When the dry season begins, the grass stops growing and turns brown. Most trees lose their leaves.

Thunder and lightning storms usually start near the end of the dry season.

Sometimes a spark of lightning starts a fire in the dry grass. The fire spreads over many square miles. Only the most hardy types of grasses, trees, and shrubs can survive. But the fires also help new plants to grow. The burning clears away dead plants and adds valuable minerals to the soil.

Soon the rains return. They put the fires out. The grasses send up new shoots, and new buds poke out of the tree branches. Life comes back to the savannah.

3

Grasses and other plants start the chain of life on the savannah. Herds of wildebeests (WILL-duh-beests), zebras, and other **grazers** move about looking for food. Grazers are animals that eat mostly grass.

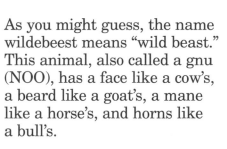

As you might guess, the name wildebeest means "wild beast." This animal, also called a gnu (NOO), has a face like a cow's, a beard like a goat's, a mane like a horse's, and horns like a bull's.

Small bands of zebras often mix with the wildebeests. The zebra is sometimes called a horse with stripes. No two patterns of stripes are exactly the same. When the zebras crowd together in a herd, the many stripes make it hard for enemies to spot or attack a single animal.

Impalas are graceful, gentle plant-eaters that also gather in herds. Like many other grazers that live on the savannah, they have great running and jumping abilities. When frightened, an impala leaps high into the air. This may startle intruders. The impala can also run from danger at speeds as high as 50 miles per hour.

Many grazing animals do not stay in the same grasslands all year long. Instead, they **migrate,** or move, from one place to another. The wildebeests and zebras of eastern Africa spend the rainy season on the open grassy plains of Tanzania (tan-zah-NEE-ah). But when the rains end and the grasses begin to die, giant herds head north to Kenya in search of food and water.

The migrating herds finally arrive in places where streams and rivers flow. The waters help the tall grasses that the animals feed on to survive. The rivers and streams also provide drinking water for the thirsty herds during the long dry season.

As the rainy season approaches, the herds begin the return trip to the open plain. The river crossing they must make is the most dangerous part of the journey.

Sometimes the river is swollen. Drowning in the rushing water is one danger they face. Being eaten by hungry crocodiles is another. Some animals will not survive the great migration, but most do. In a few weeks they will reach their rainy-season home, the same grasslands that they left many months before.

African elephants eat some grasses, but more often they eat the leaves and twigs of trees and shrubs. Animals that feed mostly on leaves and twigs are called **browsers**. Elephants are browsers that roam long distances over the African savannah. They are the world's largest land animals.

When they are thirsty, elephants sometimes dig into the earth with their tusks and trunks to reach water. Later, other animals come to drink at the water holes the elephants have made.

Another browser, the giraffe, eats the high leaves of trees that other animals can't reach. Its short skin-covered horns help keep thorny branches out of its eyes.

Giraffes have sharp vision and good hearing. And because they are so tall, they can see enemies before other animals can. In fact, other savannah animals often keep safe by following the giraffes. As soon as the giraffes start to run, the other animals flee too.

Small herds of two-horned rhinoceroses live on the African savannah. Birds called oxpeckers hitch rides on the rhinos' backs. The little birds eat ticks and other pests that they find living on the rhinos' skin.

There are two different species of rhinos in Africa: black rhinos and white rhinos. Surprisingly, the main difference between the two isn't in their color but in their lips. The white rhino—which is a grazer—has flat, square lips. The black rhino has pointy, flexible lips, which it uses to browse on leafy twigs and shrubs.

Baboons are one of the few kinds of monkeys that live on the savannah. They eat many different kinds of foods, including grasses, leaves, roots, birds' eggs, crocodile eggs, insects, and scorpions.

Baboons live together in groups called troops. Each troop is made up of about 50 animals. When baboons are not eating or looking for food, they can usually be found grooming each other and their young.

Some animals eat the grazing and browsing animals of the savannah. The animals that are eaten are called **prey**. The animals that do the eating are called **predators**. Predators must hunt in order to live. The largest predator on the savannah is the lion.

Lions live in large family groups, or prides, of as many as 30 members. The female lions do most of the hunting.

Once the prey is caught and killed, it is shared by all the members of the pride. The male lion usually eats before the females and cubs do. A large meal like this wildebeest keeps the big cats satisfied for three or four days.

Leopards and cheetahs are smaller relatives of the lion. Each of these graceful predators has its own special strategies for hunting. Leopards rest during the day, often on a tree branch, and silently stalk their prey at night. Cheetahs, which hunt during the day, rely on their speed. They are the fastest four-legged animals in the world.

Leopards are not only stealthy—they are also very strong. A leopard can drag the carcass of a large animal like an impala up into a tree and wedge it between branches. It then eats as much as it wants and leaves the rest for the next day or two.

Leopards and cheetahs both have spots. But only cheetahs have black lines on their faces.

A cheetah can run as fast as 70 miles an hour when chasing its prey. But it can keep up that speed for only a short time before getting tired.

Other animals that live on the savannah belong to the clean-up crew. These animals, known as **scavengers**, eat the predators' leftovers. Hyenas and vultures are two well-known scavengers. They patrol the grasslands looking for animal remains.

The jackal, a kind of wild dog, is another scavenger that roams the savannah. Like hyenas, jackals do not live by scavenging alone. Both animals also hunt small and medium-sized prey for food.

Insects, worms, and other **decomposers** are also part of life on the savannah. These small but important creatures return nutrients to the soil by eating and breaking down dead plant and animal matter.

Termites eat huge amounts of plant matter. They also make tunnels underground, which loosens the soil and helps grasses grow.

Among all the decomposers on the savannah, termites are the easiest to spot. From soil, they build tall nests held together with their saliva. The nests often take fantastic shapes, and some are taller than an adult human being.

Many different savannah animals find their own uses for termite mounds. Cheetahs, lions, and other predators often rest on top of the mounds or use them as lookout posts for spotting prey. Meanwhile, plant-eating animals, such as these topis (TOH-peez), use them as lookouts for spotting predators.

Many of the earth's natural grassland habitats have disappeared, replaced by big ranches, farms, and cities. But with the help of people in many African countries, large areas of the savannah have survived. The land is still covered with tall grasses and dotted with trees. And it is still home to some of the largest, fastest, and most fascinating animals in the world.